DATE DUE			
GAYLORD			PRINTED IN U.S.A.

HUBBLE SPACE TELESCOPE

Published by Smart Apple Media

123 South Broad Street

Mankato, Minnesota 56001

Photos: page 5—NASA/STScI/Brack Whitmore; page 7—Corbis;

page 9—NASA/Kennedy Space Center; page 13—NASA/Johnson Space Center;

page 15—NASA/NSSDC; page 17—NASA/Kennedy Space Center;

page 19—NASA/JPL/Caltech/J.T. Tranger; page 20-21—NASA/JPL/Caltech; page 23

left—NASA/JPL/Caltech/Raghvendra Sahai & John Trauger, WFPC2 team; page 23

right—NASA/STScI; page 24—NASA/NSSDC/A. Coulet (ST-ECF,ESA); page 25—NASA/NSSDC/

C.R. O'Dell, Rice Univ.; page 26-27—NASA/NSSDC/Jeff Hester & Paul Scowen,

Ariz. State Univ.; page 29—NASA/STScI; page 31—NASA/STScI

Design and Production: EvansDay Design

Library of Congress Cataloging-in-Publication Data

Hakkila, Jon Eric, 1957–

Hubble Space Telescope / by Jon Hakkila and Adele D. Richardson

p. cm. — (Above and beyond)

Includes index.

Summary: Describes the Hubble Space Telescope and its

role in space exploration.

ISBN 1-58340-047-8

1. Hubble Space Telescope (Spacecraft)—Juvenile literature.

2. Outer space—Exploration—Juvenile literature. [1. Hubble Space Telescope

(Spacecraft). 2. Outer space—Exploration.]

I. Richardson, Adele, 1966–. II. Title.

III. Series: Above and beyond (Mankato, Minn.)

QB500.268.H35   1999

522'.2919—DC21   98-19948

First edition

1 3 5 7 9 8 6 4 2

HUBBLE SPACE TELESCOPE

DR. JON HAKKILA & ADELE D. RICHARDSON

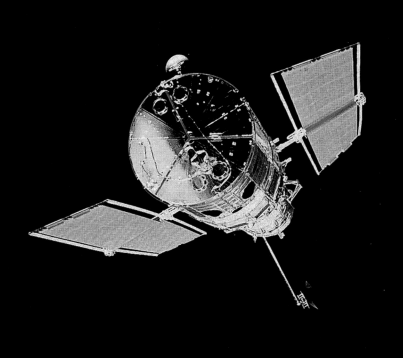

The *Hubble Space Telescope* circled high above the earth, its cameras and instruments pointed at the planet Neptune ✳ Suddenly, it found a strange development ✳ The information was quickly sent to a nearby communications satellite, which relayed it down to the Goddard Space Flight Center in Maryland ✳ Computers processed the information and passed it on to the nearby Space Telescope Operations Control Center ✳ There, astronomers and scientists huddled around a computer screen to catch a glimpse of *Hubble*'s find ✳ It seemed the temperature of Triton, a moon of Neptune, was rising ✳

The Dream Comes True

The idea of placing a telescope in orbit around the earth was born long before the rockets needed for launch were invented. During the 1920s, a German scientist named Hermann Oberth believed that rockets powerful enough to lift a massive telescope into orbit would someday become a reality.

Twenty years later, the German V-2 rockets were invented. These rockets, which were the largest ever built at the time, offered a small bit of hope for Oberth's dream. Many of the German scientists that created the V-2 later became American citizens. During the 1950s, they joined the United States' space program, bringing their knowledge and ideas with them.

Some **astronomers** in the U.S. also believed that a telescope could be put into space. Lyman Spitzer, an astronomer at Princeton University, first suggested the notion in 1946. In 1970, he and other astronomers presented the idea to the National Aeronautics and Space Administration (NASA). They argued that the space program had already launched hundreds of **satellites** and had

A German V-2 rocket being prepared for launch.

even landed men on the moon. After such achievements, the astronomers argued, NASA should certainly be able to place a telescope in orbit.

NASA agreed to undertake the project, but there were a few problems to overcome. The first, and largest, of these problems was a lack of money. Billions of dollars

An **astronomer** is a person who studies celestial objects such as planets and stars.

A **satellite** is an object—natural or man-made—that orbits a celestial body.

had been spent on moon landings, and the U.S. government was reluctant to give NASA the money needed to build and launch a space telescope.

The second obstacle was the feelings of the American public. After the moon was conquered, many people lost interest in the space program and felt that the money would be better spent in studying Earth. Many people also wondered how a telescope would be repaired if something went wrong. With all of these obstacles and unanswered questions, the astronomers' dream seemed unlikely for several years.

Probably the greatest reason the space telescope did become a reality was the space shuttle. This new spacecraft, which resembled an airplane in build and operation, provided the space program with a way to place a telescope and satellites into space and to retrieve and repair them. The first shuttle, *Enterprise*, was completed in September 1976.

The telescope, then called the Large Space Telescope, was designed by NASA in 1979. The original plan was to return it to Earth every five years, repair or replace whatever was needed, and then relaunch it back into orbit. However, since the shuttle made repair in space possible, NASA decided that four in-orbit servicing missions would be enough to maintain the telescope. The first two missions, in December 1993 and February 1997, were both successful. The other two missions, scheduled for 1999

A space shuttle, NASA's means of placing Hubble in space.

and 2002, involve instrument upgrading and the repair of any damaged parts.

The space telescope that was just an idea in the 1920s is today called the *Hubble Space Telescope.* It was named after Edwin Powell Hubble (1889–1953), an American astronomer who made some of the greatest discoveries about our universe. He learned that the universe contains many star systems like our own **Milky Way** Galaxy; he also was among the first astronomers to find evidence that the universe is growing.

Hubble was launched on board the shuttle *Discovery* on April 24, 1990. To place the telescope into orbit, *Discovery* had to fly about 380 miles (610 km) above the surface of the earth. To date, this is the highest altitude a shuttle has ever flown. This would be only the beginning of many "firsts" involving the *Hubble Space Telescope.*

*The **Milky Way** is the galaxy in which we live.*

Since released into orbit, Hubble has opened many windows to the universe.

The Space Telescope

Controlling *Hubble* in orbit is done by radio commands from the Goddard Space Flight Center in Greenbelt, Maryland. However, all of the scientific observations are coordinated with and conducted through the Space Telescope Science Institute in nearby Baltimore, Maryland. The Institute was actually created specifically for the telescope. The European Space Agency (ESA) also helps NASA with the operation of the space telescope program and has built many of the telescope's parts.

Hubble weighs about 25,000 pounds (11,250 kg). It looks like a giant tube with two wing-like solar panels on either side. These solar panels provide the telescope's power; as sunlight hits the panels, it is changed into electricity, which is then stored in six large batteries on board.

Hubble is a reflecting telescope, which means that it uses mirrors—rather than a lens—to collect and focus light. The telescope was designed to detect specific types of light within the **electromagnetic spectrum**. The light that *Hubble* sees is only the part of the spectrum called visible light: the light that people can see with their eyes.

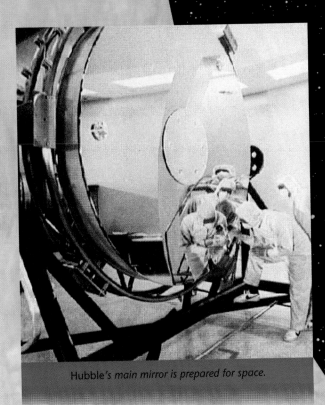

Hubble's *main mirror is prepared for space.*

Hubble's main light-collecting mirror is nearly eight feet (2.4 m) across. This makes Hubble a large telescope, but it is still much smaller than those found on Earth, which can be as large as 49 feet (15 m) across. The light that Hubble detects is collected with the main mirror, passed on to a smaller mirror, and then sent to other instruments that photograph it and determine what kind of light it is.

The **electromagnetic spectrum**
is a range of wavelengths.

APRIL 1990

Shortly after Hubble's orbit begins, a problem is found with its main mirror.

Telescopes on Earth have several limitations that do not affect *Hubble*. Earth's atmosphere prevents large amounts of light from reaching the ground. The atmosphere also contains moving air, which causes distant images to look fuzzy; this is what causes stars to twinkle. *Hubble*, placed high above the atmosphere, is free of these problems. It can gather more kinds of light and has no moving air, smog, or bright lights between it and the images it is studying. As a result, the pictures *Hubble* sends to astronomers are 50 times clearer than those provided by telescopes on the ground.

Along with its mirrors, *Hubble* uses four main instruments to record images in space. Two of these are cameras. The Faint Object Camera takes pictures of tiny, hard-to-see objects in space, while the Wide Field/Planetary Camera 2 takes pictures of larger objects.

During a mission by the shuttle *Endeavour* in December 1993, the crew replaced the telescope's old Planetary Camera with Camera 2. Because the main mirror was found to be faulty soon after launch, the *Endeavour*'s crew also installed corrective optics. This equipment works much the same way that glasses help correct a person's vision. *Hubble*'s pictures were at first blurry and out of

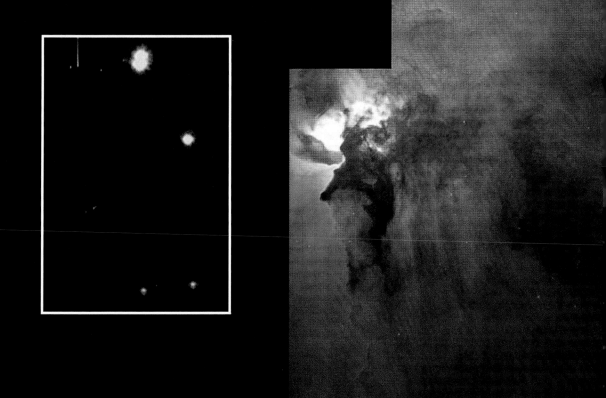

Hubble's first picture (left) with the faulty mirror; a picture of the Lagoon Nebula (right) after corrective optics were installed.

focus, but since the corrective optics were installed, the telescope's image clarity has been spectacular.

The other two main instruments are **spectrographs**. These instruments view light and separate it into its different wavelengths. The different colors can tell scientists an object's temperature; its speed of travel; and its density, or how heavy it is. *Hubble*'s spectrographs are used to

A **spectrograph** is an instrument used to separate light into different wavelengths.

view extremely faint, distant objects in space, as well as stars closer to our own **solar system**.

During the repair mission in February 1997, the *Discovery* crew replaced the telescope's old spectrographs with new ones. The crew also found that some of the insulation was going bad. They solved this problem by attaching insulated blankets to the telescope.

Hubble circles the earth every 95 minutes, but all of its time isn't spent working. Occasionally, the telescope takes a "housekeeping" break. During this time, *Hubble*

Hubble *is released in space (left), and the solar panels are extended (right) to begin the telescope's operation.*

receives commands from Earth, sends down the data it has collected, or is repositioned to look at another object. Astronomers are careful to never point the telescope at the sun or moon, however; the brightness of these objects would damage *Hubble*'s sensitive equipment. For this reason, the telescope is constantly looking out into space.

*A **solar system** is a star and the celestial bodies that orbit it.*

In-orbit servicing missions repair damaged parts and keep Hubble *up and running.*

Exploring Our Solar System

Our own solar system is made up of the sun and the nine planets surrounding it. People throughout history have been able to look into the night sky and observe five of these planets: Mercury, Venus, Mars, Jupiter, and Saturn. But little was really known about them until scientists launched man-made satellites designed for space exploration. Since the 1960s, this achievement has allowed scientists to make numerous monumental discoveries about our solar system. In the last several years, *Hubble* has added considerably to this knowledge.

On March 7, 1996, *Hubble* gave astronomers the first clear pictures of our solar system's most distant planet, Pluto, and its moon, Charon. Pluto is not only extremely far from the sun, but it's also the smallest of the planets. Yet *Hubble* was able to photograph features on Pluto's surface, including polar ice caps.

Some of *Hubble*'s most impressive pictures have been of Jupiter, Saturn, Uranus, and Neptune. Before the telescope was in orbit, photographs of these planets had been taken by space exploration vehicles, such as *Voyager*,

Aurora, or streams of light, are visible on Saturn's atmosphere.

Pioneer, and *Galileo,* on their way past the planets. These pictures, however, only captured images of the planets at one point in time. With the *Hubble Space Telescope,* astronomers and scientists can constantly monitor these planets and watch for changing conditions, such as storms. The telescope has also taken detailed pictures of Mars, allowing scientists to see the large dust storms that regularly churn across its surface.

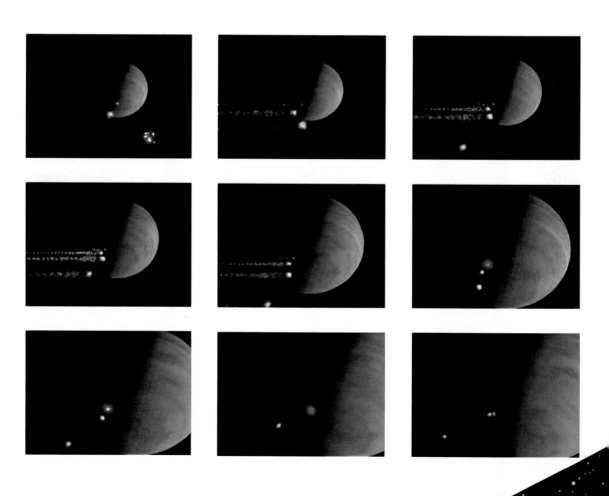

Perhaps the most startling pictures from our solar system were taken in July 1994, when the comet Shoemaker-Levy 9 slammed into Jupiter. These pictures show the horrible effects that falling space material, such as comets and meteors, can have on a planet and its atmosphere. Many scientists believe that an impact similar to the one on Jupiter may have brought about the extinction of dinosaurs on Earth millions of years ago.

In June 1998, *Hubble* looked for the first time at Neptune and discovered that one of the planet's moons, Triton, was heating up. Scientists are not sure why this is happening, or what it may mean, but the space telescope will continue to watch for further changes.

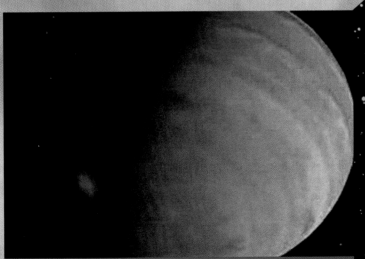

Hubble *images showing the path taken by comet Shoemaker-Levy 9 (opposite) to collide with Jupiter (above).*

Star Gazing

Stars are giant spheres of heated gas. All stars are far from Earth, which is why they look so small when compared to the sun. The closest star other than the sun is Alpha Centauri, which is about four **light-years** away. A light-year is the distance light can travel in one year—about 5.8 trillion miles (9.46 trillion km).

Other stars are much farther away. At such great distances, even the powerful *Hubble* cannot make stars look any larger than a bright point of light. That's why the telescope's spectrographs are so important. Even with only a dot of light, the instruments can separate the light and determine the star's various properties: what it's made of, how hot it is, and how fast it's traveling, for example.

Hubble has been able to take pictures of some extremely faint stars that have never been directly seen before. Some of these are brown dwarf stars, which are cool, dim stars that contain too little material to burn like a true star, yet are hotter and brighter than planets. In the past, brown dwarf stars have been difficult for astronomers to detect, but *Hubble* has had no trouble finding them.

Hubble has also photographed a **neutron star**. This is a massive star that has used up its fuel, and its outer layers have exploded in a **supernova**. The center of the star doesn't explode, however, because gravity holds it together. In fact, after the outer layers have blown off, the center contracts, or becomes smaller. As gravity contracts the inner material of the center, a neutron star can become very small but heavy. Scientists call such stars

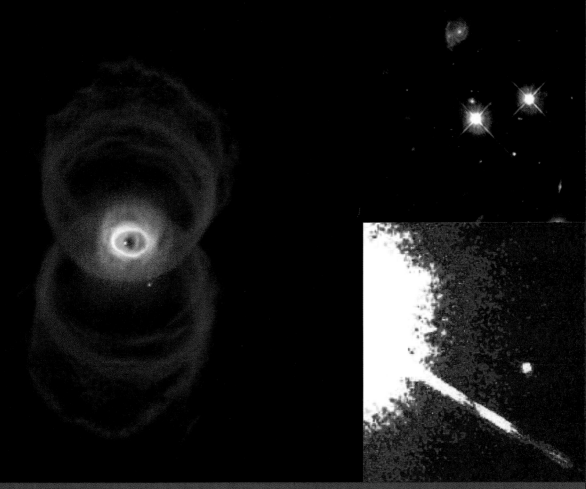

Clockwise from left: an eerie photograph of the Hourglass Nebula; Hubble's 100,000th observation; a spectacular brown dwarf star.

"dense." Despite their small size, *Hubble* is able to identify and photograph neutron stars for astronomers to study.

When a neutron star is still young, it spins around rapidly and beams particles of matter and radiation into space. To an astronomer looking through a telescope, these beams are flashes, or pulses, of visible light. These objects are appropriately named **pulsars**. Many astronomers have viewed pulsars from Earth-bound tele-

*A **pulsar** is a spinning star that sends out bursts of radio signals.*

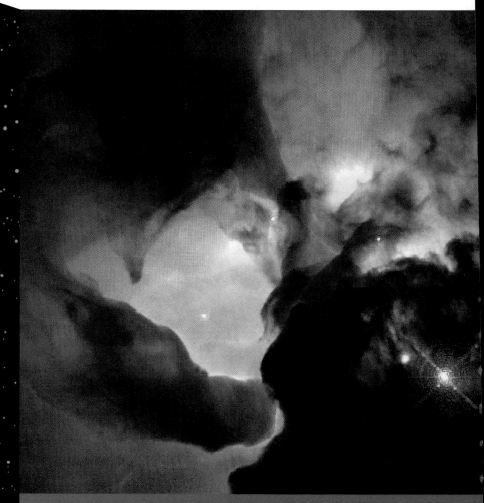

This clear image of the Lagoon Nebula demonstrates the power of Hubble's cameras.

scopes, but *Hubble* has offered clearer pictures—and more of them—than anything previously seen.

Some of *Hubble*'s most interesting observations have been of a space phenomena called **nebulas**. The word "nebula" is a general term used to describe objects that look like clouds in space. Some of the largest nebulas, which can be seen through a small home telescope, are giant gas and dust clouds from which stars are formed. Some

The haunting beauty of the Great Orion Nebula is captured by the space telescope.

25

of the nebulas that have been discovered are the Crab Nebula, the Great Orion Nebula, and the Eagle Nebula.

Hubble has studied the Crab Nebula in detail. In the year 1054, a supernova explosion took place where the nebula now exists; this expanding shell of hot, glowing gas is all that is left of the exploded star. *Hubble* has been able to watch the expanding gas cloud and has spotted a pulsar in the middle of it.

The telescope has also taken impressive pictures of the Eagle Nebula. This growing gas and dust cloud is currently the birthplace of many new stars. A star forms when gravity pulls together the gas and dust into part of a cloud. As the particles fall faster toward the center, they heat up and eventually become so hot that the forming star begins to shine. Usually, astronomers cannot see forming stars because all of the falling material creates a cover, or cocoon, around the new star. The stars will only become visible once enough heat is generated to blow off the cocoon. There are a few new stars in the Eagle Nebula

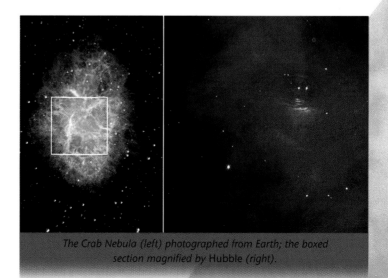

The Crab Nebula (left) photographed from Earth; the boxed
section magnified by Hubble (right).

The Eagle Nebula and its unique pillar formations (inset).

that are shining brightly enough to blow cocoons off of their star neighbors.

Hubble has also been able to study the remains of a recent supernova. Astronomers saw the star explode in 1987 and have been watching the gas cloud expand ever since the telescope was placed in orbit. The supernova is so distant that only Hubble is able to take pictures of it.

Another interesting and energetic space phenomena observed by Hubble are **black holes**. Using the laws of physics, scientists believe that black holes are formed

A **black hole** is a former star that has been crushed by extreme gravity.

from the centers of stars that collapse under an explosion. As a star collapses, its gravity becomes so strong that it is crushed into a small, dense core. The force of the gravity that surrounds such an object is equal to the speed of light, meaning that the star retains its light and appears to be black.

Although black holes do not give off any light, they are actually far brighter than any normal star. This is due to a process called **accretion** that occurs when materials— such as meteors, asteroids, and dust—drawn toward a black hole's surface collide with one another and produce light energy.

Scientists believe that black holes in the center of a galaxy can gradually combine and form gigantic black holes. *Hubble* has observed the accretion surrounding some of these holes. The light given off is incredibly bright; some of these black holes are believed to pull in amounts of material equivalent to several billion of our suns!

Accretion *is a process in which colliding objects release energy and give off light.*

Accretion surrounding a massive black hole (left) is made even brighter when magnified (right).

Focusing on the Future

Although *Hubble* has already accomplished much, it still has a lot of work to do. Unfortunately, many research projects have been postponed or rejected altogether because there simply has not been enough telescope time available to astronomers.

The good news for astronomers is that NASA has decided to extend the mission until 2010. During the final *Hubble* servicing mission in 2002, many of the systems on board will be upgraded or replaced entirely with new state-of-the-art technology, giving astronomers a better view of our expanding universe.

In 2007, a sister telescope, called the *Next Generation Space Telescope*, will be launched to work alongside *Hubble*. The *Next Generation* telescope will have a main mirror twice as large as *Hubble*'s, allowing it to gather twice as much light. The new telescope's orbit will also be set much higher, and its spectrographs will study different wavelengths of light. Together, the combined capabilities of the two space telescopes should greatly widen the range of scientists' space exploration. When the *Hubble* mission ends in 2010, a space shuttle may be used to

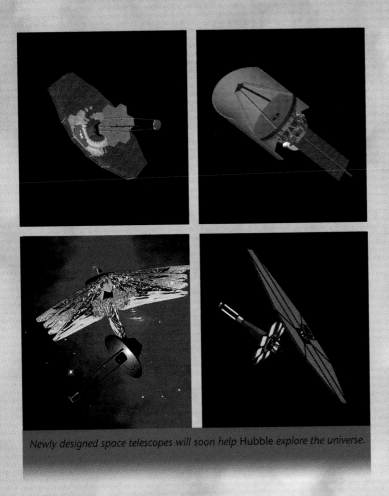

Newly designed space telescopes will soon help Hubble *explore the universe.*

bring the telescope back to Earth, or astronomers may continue to use it until it breaks down.

Over the course of the last decade, the *Hubble Space Telescope* has brought astronomers closer to the stars, opening the doors to some of the universe's most amazing phenomena. Its discoveries, both on Earth's neighboring planets and in galaxies light-years away, will continue to expand our knowledge of a changing universe.

INDEX